WARRIORS

HARPER PHOTOGRAPHERS

PHOTOGRAPHS BY

RICHARD STACK

WARRIORS

HARPER COLOPHON BOOKS
Harper & Row, Publishers
New York Evanston San Francisco London

To Anh and Justin

First HARPER COLOPHON edition published 1975.

LIBRARY OF CONGRESS CATALOG CARD NUMBER: 75-18632

STANDARD BOOK NUMBER: 06-090319-8

Designed by Patricia Dunbar *A Black Star Book*

75 76 77 3 2 1

ACKNOWLEDGMENTS

There are far more people to be thanked than I could ever list here, but I do want to note my gratitude to a special few.

To Hiroshi Isogai and Shunjiro Matsushimu of Group Pam Tokyo for having gotten me interested in photography in the first place.

To Howard Chapnick, who took time from his responsibilities as president of Black Star to work on the planning and making of this book. His patience and his confidence in me are deeply appreciated.

To all the other people at Black Star, who helped in dozens of little ways and a great many big ones as well.

To Jack Giacopelli and Werner Wolff of G&W photographic service, who put up with my lengthy absences during the shooting of this book. Also to Lou Alexander of G&W, a good friend and a great printer; his magnificent prints have been used in preparation of this book.

To Nach Waxman, my editor at Harper & Row, for assembling the text of the book from my tape and interview materials and for his helpful suggestions at every step of the way.

And finally, a special acknowledgment to the officers and men of Parris Island. Their cooperation, encouragement, and good humor were all that I could have asked for.

A NOTE ON TEXT SOURCES

There are four different types of text accompanying the photographs in this book.

First, there are replies by recruits to a questionnaire that I handed out to them toward the end of their training period. These are designated *Responses to a written questionnaire.*

Second, there are bits of conversation involving recruits or drill instructors that I taped on the spot in the barracks or mess hall or in the field. This material is called *Taped dialogue.*

Third, there are interviews that I conducted with various drill instructors. Material from this source is noted as *Taped interview.*

Fourth, there are a number of quotations taken from the Marine Corps basic manual of instruction, a copy of which is given to every recruit entering Parris Island. This manual, officially designated *The Marine Notebook,* is more familiarly called "The Red Book." Material taken from it is followed by the reference *The Red Book.*

It will probably be apparent that the material offered in the text was not necessarily gathered in conjunction with the shooting of any particular photograph, and there is no suggestion made that the individual or individuals shown in a given picture said or wrote what appears in the adjoining text. All quotes, however, were actually gathered by me and appear as they were recorded or taken off the questionnaires.

THE BACKGROUND

Properly known as the Marine Corps Recruit Depot, Parris Island, South Carolina, it is the most important of the few places in the country where the United States Marine Corps takes boys and makes them into men. It is guaranteed that no one who has spent time there will ever forget the place.

There are a few basic physical facts to know. Parris Island lies approxmately eighty miles south of Charleston, South Carolina, forty-five miles north of Savanna, Georgia, and two-and-a-half miles from the city of Beaufort, South Carolina. It really is an island. The *Marine Notebook* given to every new recruit makes that inhospitable fact abundantly clear. Parris Island, it notes, with some satisfaction,

> is surrounded by marsh (swamps) and tidal waters. What may appear to be firm ground at low tide may be seven or eight feet under water at high tide. There is also quicksand in various places. These tidal areas are the natural habitat for poisonous snakes. In addition, due to the tides, strong eddies and currents make swimming in these waters extremely difficult even for the most experienced swimmers.

The island's only connection with the mainland is a barren man-made causeway, lined with palmettoes and scrub going right up to the main gate, where a huge sign in traditional gold and red says WELCOME TO THE UNITED STATES MARINE CORPS RECRUIT TRAINING DEPOT, PARRIS ISLAND, SOUTH CAROLINA. At night the sign is illuminated by four stark white high-intensity floodlights, the first sign of life the traveler has seen for miles. Most travelers on the causeway, though, are not very interested in the scenery along the way. They are teenage boys, eighty or so at a time, coming in from their pickup points on the big, gray Marine Corps buses, and their only interest is in what's behind the gate.

Entering Parris Island as a recruit is like entering a special kind of cloister. From the outset the recruit is conscientiously separated from his past. The isolation is almost complete. No radio, no television, no Coca-Cola, no hamburgers, no beer, no telephone, no girl friend, no street corners, nothing that is familiar. The induction process of boot camp lasts twenty-four hours a day, seven days a week, for eleven weeks. From the moment the newcomer steps off the bus and finds himself standing on painted yellow footprints in the dead of night, staring blankly at a man in a funny looking Smokey the Bear hat, he learns he is a recruit private in the United States Marine Corps.

"You people get off here," barks Smokey the Bear. "Walk down those steps to those yellow footprints across the way. You *will not* move till I tell you to. Now, when I say so, you *will* get yourselves on those yellow footprints in columns of threes. You three, right next to each other—get off the bus. NOW! Get over there. Let's go! Move like you're alive. Right over there where you see those double footprints. Move it. MOVE, MOVE, MOVE!! OK, face this way. Eyes straight ahead. STRAIGHT ahead. When I give the word, you *will* get yourselves inside this building. Now inside this building there are numbered tables. And when you go inside the building you *will* place your right hand on the number in front of you. From one to thirty-four. Your right hand on the number only. You *will not* place your gear on those tables. I just want your right hand on that number. Is that clear?"

Once inside the Recruit Receiving Barracks the new arrival is shown how to stand and how to sit and how to do both on command:

"Alright now. First thing we teach you here is the position of attention. *You.* Stand up on the table. The first thing we're going to concentrate on is your feet. Get your feet to form an equal angle of 45-degrees, heels on the line and touching in the rear. Hand: thumbs on the trouser seam, palm against your leg. Head and eyes straight ahead. Tuck the chin in. Roll the shoulders back. Suck the gut in. And you *will* keep your mouth shut.

"And when I tell you to sit down you'll sit down right on your ass right where you're at. If you do so much as *move* a muscle, you'll be standing right back up at attention. Do you understand?"

"Yes, Sir!"

"Sit down!"

"You're slow—get up!"

"GET UP!"

"Now we'll try that one more time. When I say sit down, all of your asses better be hitting that deck in one big swish. SIT DOWN"

"GET UP!"

The recruits jump to attention looking straight ahead, their eyes open so wide they practically pop. Everyone is searched for contraband: knives, any sharp object, alcohol, drugs and shooting-up equipment, pictures larger than wallet-size, etc., etc., etc. Then they are given a 45-second haircut, marched from there to Hygienics, stripped of their clothing, given a one-minute shower, and ordered to shave. They are quickly and impersonally photographed, inspected, injected, and fingerprinted. Their civilian clothing is wrapped in stiff brown paper and tied with a cord, ready to be mailed back to wherever they've brought it from.

No small matter, those clothes. They were the last link, and now the recruits stand in a line, nude, cold, and scared. Two hours before they had been standing in civilian clothes, thinking civilian thoughts. Now as green, fresh, scared recruits, they think nothing. The vacuum is almost complete and the remolding can begin.

Each recruit receives three sets of olive green utilities (shirt and trousers), two pairs of boots, and a too big green hat, which usually drops down over his ears. Emerging from Hygienics, dressed after a fashion, the recruits look like a group of Fellini extras or someone's Halloween joke. But it's no joke to the men in the Smokey the Bear hats—the Drill Instructors—who have to establish their authority immediately or risk losing it for good.

Staff Sergeant Benkeirik has seen three wars and a total of some twenty years of service. He's lean and tanned, his hair shorn completely, and he stands tall, his shoes so highly shined you can see yourself in their reflection. The starch in his shirt and trousers completes the picture. His voice is piercing:

"OK now, I'm going to tell you something here and all you have to do is listen up and pay attention because it deeply concerns *you.*

"Privates, you are now in Recruit Receiving Barracks, Marine Corps Depot, Parris Island, South Carolina. This is your first step toward becoming a member of the world's finest military organization, the United States Marine Corps. Throughout your career as a Marine, always be mindful of the oath of enlistment you took when joining us and do your level best to live up to it. What does that mean? Simply this: you held your right hand up and you got yourself sworn in. You signed a legal document called a contract. It is binding and it is legal, whether it's for two, three, four, or five years. You remain with us for approximately one to two days, during which time you will be processed and formed into Recruit Training Platoons. After this you will be turned over to your own team of Drill Instructors—D.I.'s, three of them—who will train you for the next eleven weeks.

"If you came down here with the right attitude, you will graduate in eleven weeks, a man and a Marine. If you have a bum attitude or a lousy attitude or a shitty attitude, you're going to get your ass thrown in jail so fast it's going to make your head spin. They don't sympathize with you down here. They are trying to toughen you up and they got to harden you up. If you can't take it, then you can go back to your mumma. Because Parris Island is where they part the men from the boys. Remember that when you start getting weak or you start turning chicken.

"During the different phases of processing always be mindful, do what you are told when your are told, and speak only when spoken to by authorized personnel. Authorized personnel on this Island are the Marines on this Island, not that private standing on your right or left. Don't bother to ask him a question when he knows *nothing* about what's going on down here. The *only* people you are supposed to be talking to are your Instructors. Remember that. Since you are members of the Armed Forces you are under the jurisdiction of the UCMJ—that's the Uniform Code of Military Justice—passed by Congress and signed by the President in 1951 commonly known as the law of the military. A violation of any one of its articles, for example failing to obey or do what your are told when you are told, in other words disobedience of orders, could result in a trial by court martial and possible brig sentence.

"As privates in the USMC you are the lowest of all ranks. Therefore you will conduct yourself accordingly, affording all superiors proper military courtesy. Speak only when spoken to always using the word 'Sir,' 'Yes Sir!,' 'No Sir.' If you desire to ask a question you will say, 'Sir! Private, *your name,* requests permission to speak. SIR!'

"When standing you will always be at the position of attention. You will be checked on how you are standing, sitting, and how you are even sleeping, so girls get it right the first time, you'll save a lot of aggravation, and for all you little girls that don't know what aggravation is it means getting your D.I. pissed off at you. Pick up the right habits down here not the bad or wrong ones and you'll make it.

"While on this Island you will have no financial dealings with no one. And I mean no one, is that clear ladies?"

"Yessir!"

"What?"

"Yessir!"

"I can't *hear* you, girls."

"YESSIR!"

"'Alright face to your right! Pick up your towels, place them around your neck! Put your chrome domes on your head. Pick up your seabag, place it on your left shoulder. When you move out the hatchway, you will walk down the ladder well, and when your mother-fuckin' feet hit that deck, I better see your asses moving and standing on those yellow footprints. Is that clear, ladies?"

"Yessir!"

"What?"

"YESSIR!"

"Now Move! MOVE!"

As the recruits descend the stairs the most shocking twenty minutes of their entire eleven weeks will take place here in front of Receiving Barracks. For the first time they will meet their permanent Drill Instructors. These three men, highly trained, highly psyched-up, confident in their methods, move in with the swiftness of twenty and yell like a hundred; the recruit will not do anything right for the next eleven weeks.

"Boy, I don't *like* you. Get yourself on those marks, stupid! No, not those dumb dumb! *THESE!!* You are something really stupid, shitbird! You lookin' at me funny? I won't forget that, little girl, and believe me you won't either."

"You scuzzy civilians want to make me puke. You there, puke, get your ass on those footprints."

"OK now, move it!! MOVE IT!!! You're wasting my time, shitbird."

"Alright asshole to belly button, when I say move you move. You there, maggot, you want a special invitation or something?"

"No."

"No! What the fuck is No? It's No Sir, Yes Sir. Sir! Sir! Is that clear?"

"Yes Sir!"

"I'm going to *remem*ber you, sweetheart, first day here and you already pissed me off. Just wait till we get to the barn."

The seventy-five terrified and stunned recruits place their seabags on a truck and are marched a half mile to the barn (that is, the barracks, some two or three stories high containing thirty-five to forty-five double-decker bunks on each floor).

During the eleven weeks of training at Parris Island, the recruit learns that he is in a hostile environment where the individual means very little or nothing at all. His mistakes count for something, of course, and for them the group may suffer collectively. At any time, for the slightest mistake—maybe not even his—the recruit will find himself doing ice-cream and cake (bends and thrusts, which can be very exhausting), extra motherfuckers (jumping jacks), or good old push-ups, fifty or a hundred.

He will rise every day at 4:45 A.M., dress, make up his rack, clean the barn, and eat chow. By 7:00 he will be out for P.T. (physical training), which includes a half-mile to three-mile run, depending upon which week of training he is in. He does exercises that are designed to help him gain weight if he is underweight and lose weight if he's over. At the same time the exercises help roughen everyone's flabby muscles.

During Phase I, which is three weeks long, the recruit is taught a whole new set of basic skills: first aid, interior guard, M-14 mechanical training, use of the .45 pistol, sanitation and personal hygiene, moral leadership, keeping clothing and equipment in top shape and working order. He learns how to overcome his fear of height by running a Confidence Course that would test the boldness of Tarzan. The Slide for Life, a 40-foot tower with four ropes descending at a 45-degree angle, down which the recruit must crawl to the safety of land a hundred feet away. Below him is three feet of water to break his fall should he be so unlucky as to let go. He learns to wash his own clothes. He's taught Marine Corps history and tradition. He is told how the Marines have always been the best, never losing a battle; he's told that the Army's soldiers, doggies, as they are called, aren't too much on the ball. It seems to work. It seems to give him a type of pride that's hard to define.

Already bald, he receives a haircut every two weeks and sometimes once a week. He learns drill until his rifle is a perfect extension of his arm and his footsteps and those of the rest of his platoon sound like one. The Drill Instructors' ears are so trained that they can hear if a single foot hits the ground a second sooner or later than the next. The recruit's hand becomes sore from smacking the rifle to get the desired sound. His shins ache from marching.

He is taught how to kill. He practices and perfects the art, whether with the M-14 at 500 yards aiming for the twelve-inch bull's eye or on the bayonet course striking with cold steel at a wooden silhouette. He works out at the pugil stick course, where each recruit has a protective groin belt, helmet, and neck brace, and they attack each other with a heavy padded stick that acts like a rifle butt. He perfects the naked strangle hold, which can kill a man silently in a few seconds. Killing is taught him carefully and lovingly; it will be his trade. He learns to enjoy this practice and soon he wants real action.

The Marine Corps puts out approximately 300 hard, strong products a week for about $1,300 apiece; one new transport plane costs approximately $48-million. The reason the Corps can turn out men so cheaply is in part the tireless dedication of the Drill Instructors. These men will work some eighty to ninety hours a week, and a Senior Drill Instructor will put in 120 to 125 hours a week. And for this they only get $50 a month more than others of their rank.

The recruits have a tough time in Boot Camp but most do make it. And for those who have no motivation or are rebels, laggards, or Sick Bay commandos, the Marine Corps has a secret weapon that will straighten out all but the most resistant.

It's called the Motivation Platoon, and the Senior Drill Instructor can send any of his recruits there for one day "to get the civilian shit out of his head."

This is where the recruit can get a taste of the Old Corps, and some real John Wayne action.

The recruits start out on a 4½-hour march with full pack and rifle around Parris Island's 6,000 acres. It's a fast, hard-paced march that makes even well-trained muscles ache and brings out the sweat before the sun is up. Along the way, the motivation recruits stop at places like the General's office to see morning colors going up; march again; stop to listen to the base band playing the "Marine Hymn"; march again; stop to hear a talk at the brig: "If you go along with our program, you'll stay out of here, if not, we'll lock up your ass so fast it will make your head spin." Some more marching, almost to exhaustion. Then a stop at the Iwo Jima flag statue and some soft talk from an immaculately dressed officer about Corps and Country. Next a C-ration lunch, Viet-Nam style. And finally the ditch. The weary, disoriented recruits run, crawl, and slither on their stomachs and backs through some 350 yards of ditches and obstacles filled with mud and silt. Their olive green clothing becomes indistinguishable in appearance or feeling from the great mass of brown soft mud in which they are crawling. Their faces and hands look as if their blood is oozing through the skin. Their pores are caked, their faces splattered, their bodies soaked in slime. They emerge from the mud, scared, tired, lonely; they don't think they will last one more minute. But most of them will. They sit in the sun for awhile, the mud drying and baking on, forming a second skin. Then they got one more quiet pep talk and are sent back to their own platoons, muddy and exhausted, their resistance in tatters.

The recruit has seen, in one awful day, both sides, and the message has been clear enough: fight the system and you'll spend the rest of your time here running your tail off in your sweat-stained fatigues, an outcast without respect and without standing; go along with us and you can be one of us, a *United States Marine,* proud of his uniform and proud of his Corps. For most, after that experience, it's a choice hardly worth mentioning.

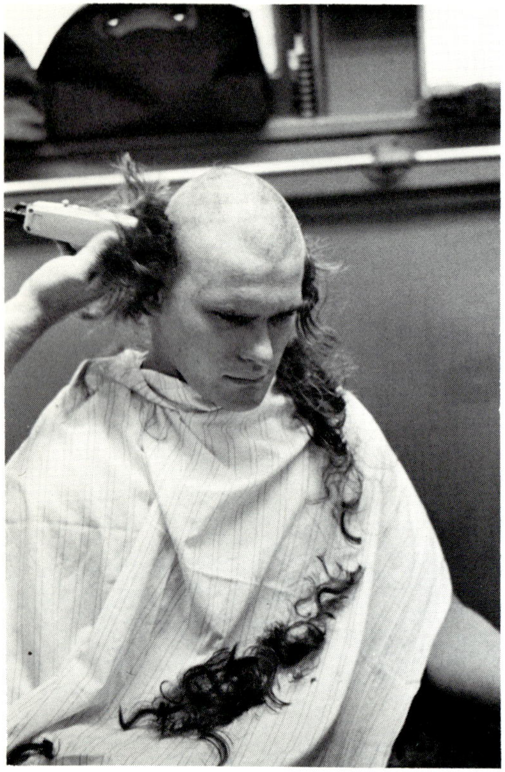

WHY DID YOU JOIN THE MARINES?

"To prove I could, and I found out it wasn't as hard as everyone said."

"I don't know."

"Prestige and discipline—experience."

"I wanted the respect a Marine gets."

"To get away from the life I had for nineteen years."

"To obtain the discipline I was missing."

"I don't know . . . to prove myself a better man."

"To become a man and respected."

"To have something in life."

"Because my brother did."

"Because of a stupid bet, and I didn't want to lose."

"To make a hard man of myself."

"To continue in a new way."

"To grow up, learn discipline, and become a man."

"Nothing better to do."

"To be a man."

"To see what kind of person I could become."

"Education."

"Because everyone said not to."

"Something I wanted to do for my country."

"There were problems concerning the support of a baby."

"For the excitement, and to see if I could make it.'

"To do something with my life."

"I was drunk out of my goddamn mind."

"I'll be a man."

—Responses to a written questionnaire

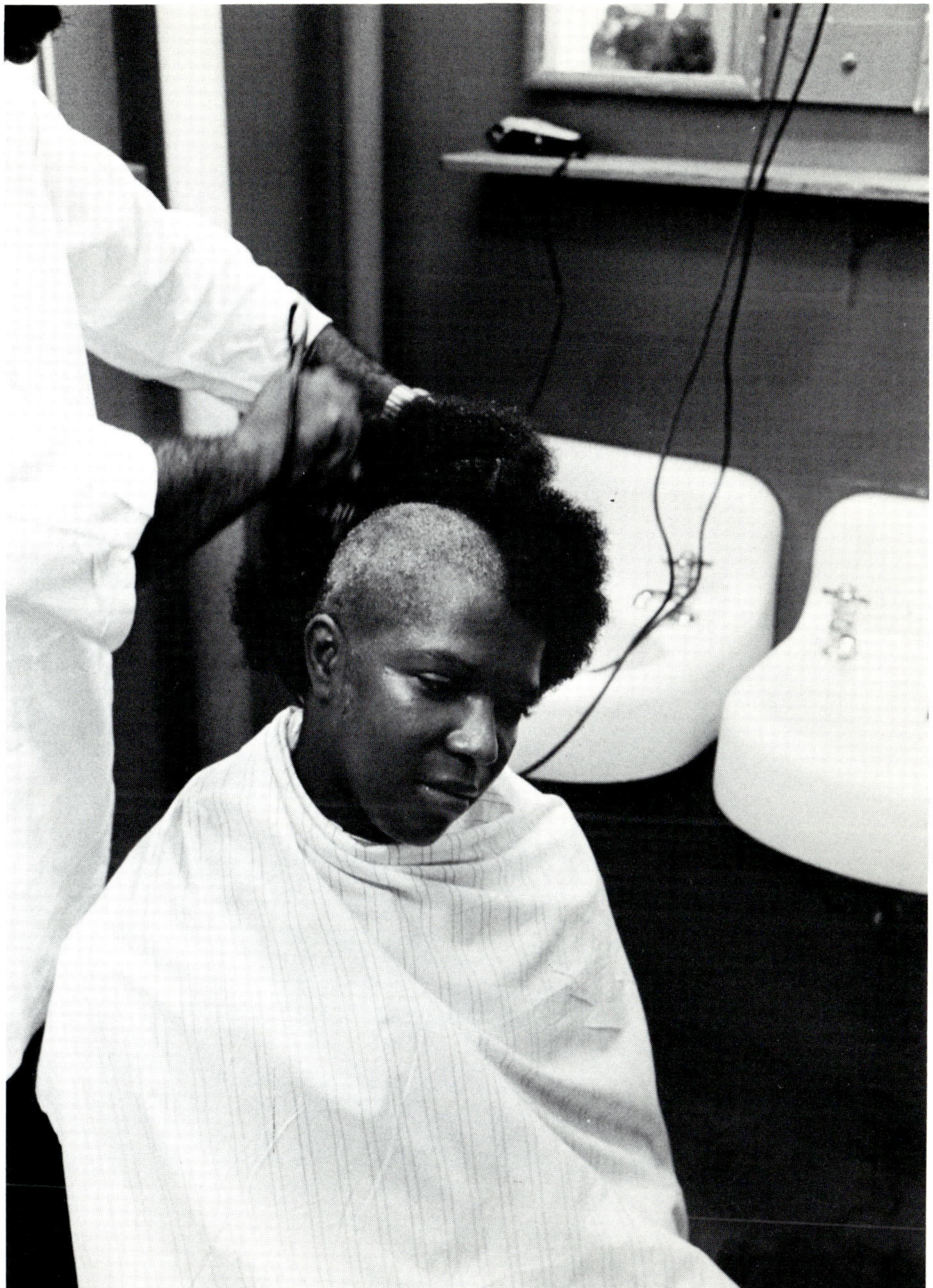

RECRUIT

WHAT DID YOU DO PRIOR TO BOOT CAMP?

"Worked on a boat in Arkansas."

"Worked in Washington, D.C., as an electrician's helper."

"Did part-time maintenance work in a nursing home."

"I just graduated from high school and had a part-time job during the first part of the summer as a stock boy. Then I quit and about a month later joined the service."

"Worked on my father's dairy farm."

"I went to work at Woodside Mills in the dye department."

"Worked construction."

"Worked as a mechanic."

"Worked in a restaurant as a cook."

"Had a part-time job as a steel worker."

"I dropped out of school at 11th grade and worked as a gardener."

"I worked at Seagel Lighting Co. putting together lighting fixtures."

"Graduated out of high school, worked in industry for 1 month, then signed up."

"I didn't finish high school. I finished the 10th grade and traveled all summer by thumb."

"Worked in a cotton mill."

"Ran pool room."

"Gravedigger at Memory Gardens Inc. I dropped out of school in 9th grade."

"Full time in a Shop Rite Foodarama Supermarket."

"Worked as full-time cook in an Italian restaurant."

"I was just fucking off around the block."

"I was a night cleaner for I.B.M. after being thrown out of school."

"Worked at Agway Grain Co., ran a front-end loader for them."

"Disc jockey at skating rink part time and machinist. Almost all my free time was spent on a bike in a bike gang."

"I worked for the Post Office as a sub-mail loader."

—Responses to a written questionnaire

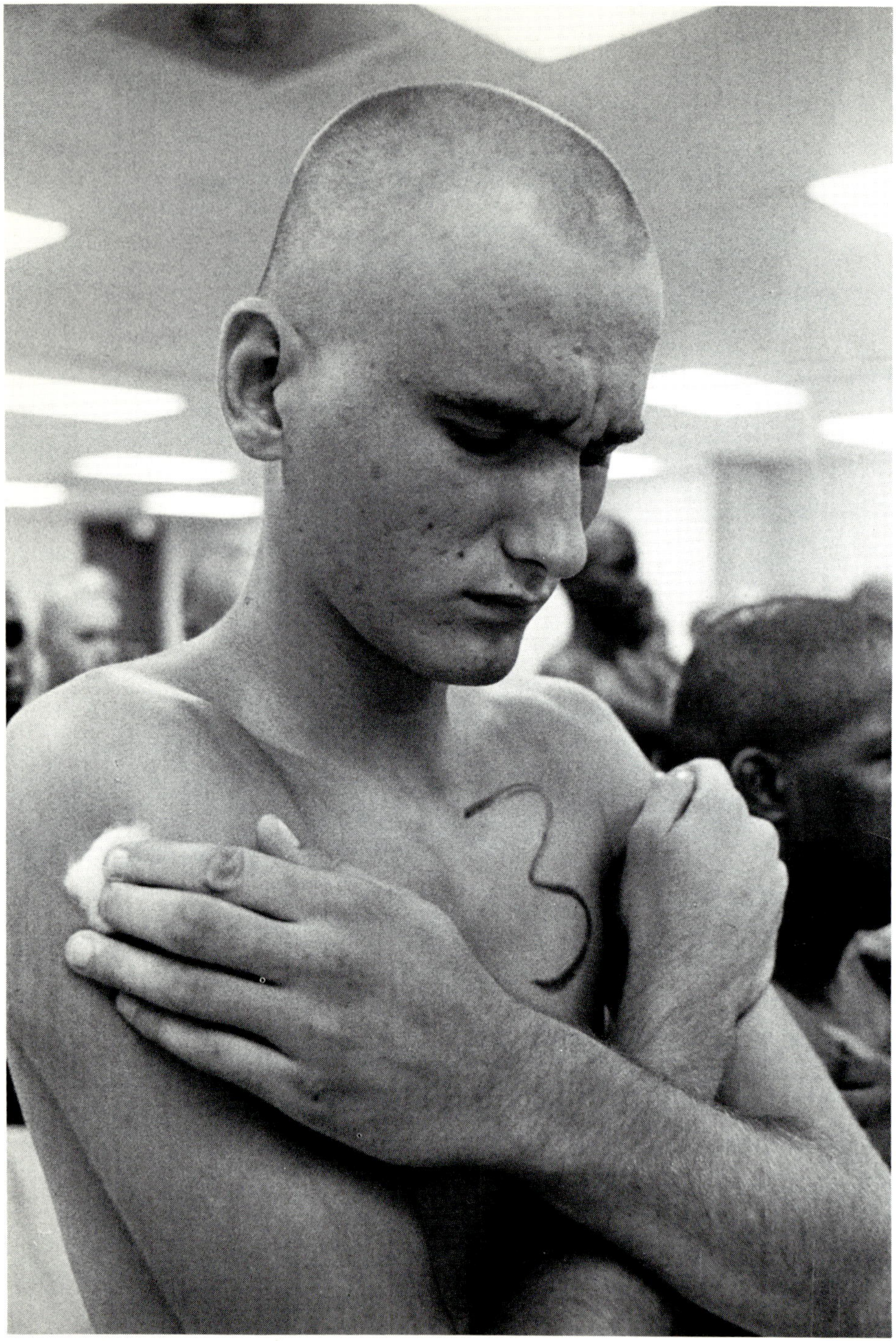

"When you get back from across the street, Private Simon . . . that *is* your name, isn't it? First name Simple, last name Simon? Huh? Is that your name?"

"No . . . Sir."

"Well, when you get back, *if* you manage to get any more goddamned information raked up out of that cesspool you call a brain, we'll make an attempt to fill out some forms."

—Taped dialogue

U S MARINE CORPS 254067

SEX HYGIENE

Venereal Disease

PREVENTIVE MEASURES: Don't get exposed. Avoid promiscuous (casual) sexual intercourse. If you do have intercourse, use a condom (rubber), urinate shortly after exposure, and thoroughly wash the exposed areas with soap and water.

IMPORTANT—If you suspect that you have a venereal disease, report to sickbay. It is NOT a court-martial offense to acquire a venereal disease. IT IS a court-martial offense if you acquire a venereal disease and *do not report* to sickbay for treatment.

—The Red Book

Your Drill Instructors are some of the Corps' finest non-commissioned officers. They are dedicated to their mission of molding you into a well-disciplined, physically fit, proficient *basic Marine.*

—The Red Book

SERGEANT B.: You know something, I think I could get out of the Marine Corps and become an actor. [Pause, laughs]

Yes, I think I could become a good actor. Because that's all we're doing down here, we're playacting 'cause we tell these privates we hate them, we call them all kinds of names, but we don't really mean it. We don't mean that they are the worst things that walk the face of the earth. We call them these names because if they should ever become prisoners of war, somebody is going to call them worse names than we call them. So if they can't take it here, they won't be able to take it over there. This brings out self-discipline.

—Taped interview

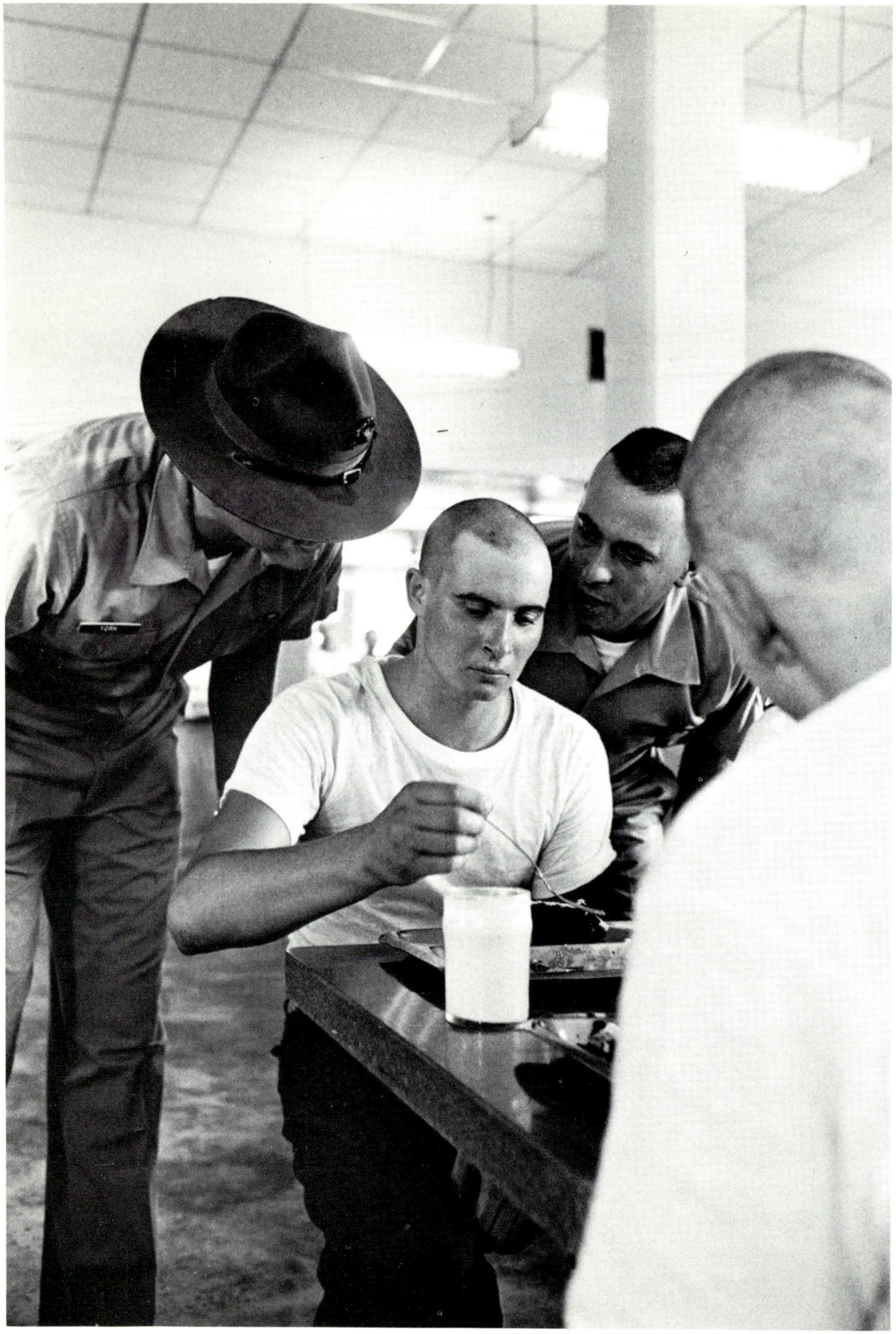

YORK

GRIFFIN

"You, Private!"
"Yes, Sir!"
"Louder!"
"Yes, SIR!"
"LOUDER!" (Drill Instructors join in unison)
"YES, SIR!"
"I'll bet you're a pussy. *Are* you a pussy?"
"NO, SIR!"
"Then you'd better sound off!"

—Taped dialogue

"Boy, I don't like you. You better move, you Communist motherfucker. I *know* the Kremlin sent you to fuck up my Marine Corps."

—Taped dialogue

RANSOM

BREITENBACH

INTERVIEWER: How many hours a day are they in actual training?

SERGEANT B.: They are training from five in the morning to nine at night. Well, not really, they have three twenty-minute chow periods and then free time from eight to nine when they can write letters and get their gear in order.

INTERVIEWER: Do you think the training is too hard for them?

SERGEANT B.: No! I think it's too easy. Compared to when you were in or when I first came in, it's much too easy. The Corps has relaxed entirely too much.

—From a taped interview

'''Number One and Two—your're going to change over. Take your left hand and face toward the woods. Take your right hand and place it between your left hand and your knees. Drop your legs, swing around, and hook over the cable. OK, One and Two, lean over the cable—LAY YOUR BELLY DOWN ON THAT CABLE! Where's Private Woods?''

''He went to the rear of the line.''

''Well, that motherfucker better be next. Number One, roll over. Number Two, roll over. Make sure that you got boot leather on that cable. Well come on, LET'S GO!''

''LAY BACK ON TOP OF THAT CABLE NOW! Yeah, smile. I'll give you something to smile about, Boy. . . . HURRY! FASTER!''

—Taped dialogue

WHAT DID YOU PROVE BY JOINING THE MARINES?

"That I could make the hardest training in the world."

"I didn't prove anything, but I found out I'm more of a man than I knew."

"I proved that I was as hard as I thought."

"I proved I was a man. No child could get through."

"I proved that I could give 100% effort, which I have never done."

"I proved that I could take anything thrown at me."

"That I was man enough to go all the way."

"I proved to myself that I had entered manhood."

"I proved I could make it with the best—something I wondered about all my life."

"That I'm as good as anyone else, if not better."

—Responses to a written questionnaire

"Private."
"Captain, Sir."
"You didn't even fucking *try* today, bitch."
"The Private did try, Sir."
"I was counting on you to fucking qualify or come close to goddamn qualifying. You didn't even fucking *try.*"
"I got mixed up . . . and I have a virus, Sir."
"You got a *virus.* You got a VIrus. No, I don't buy this shit. No fucking doctor can tell me you got a virus. Now you got no fucking duty except to go down on that fucking rifle range and fire that motherfucking rifle. Now I want you to go goddamn *qualify*! Now how the fuck can I get you to qualify if you got NO DUTY?"
"I'll go back to my duty in the morning, Sir."
"Yeah, you'll go back to duty in the MORNing. But maybe I want to fucking snap you in tonight!"

—Taped dialogue

Profanity, swearing, and verbal filth is discouraged. A person who must resort to profanity demonstrates his inability to verbally express himself in a socially acceptable fashion.

—The Red Book

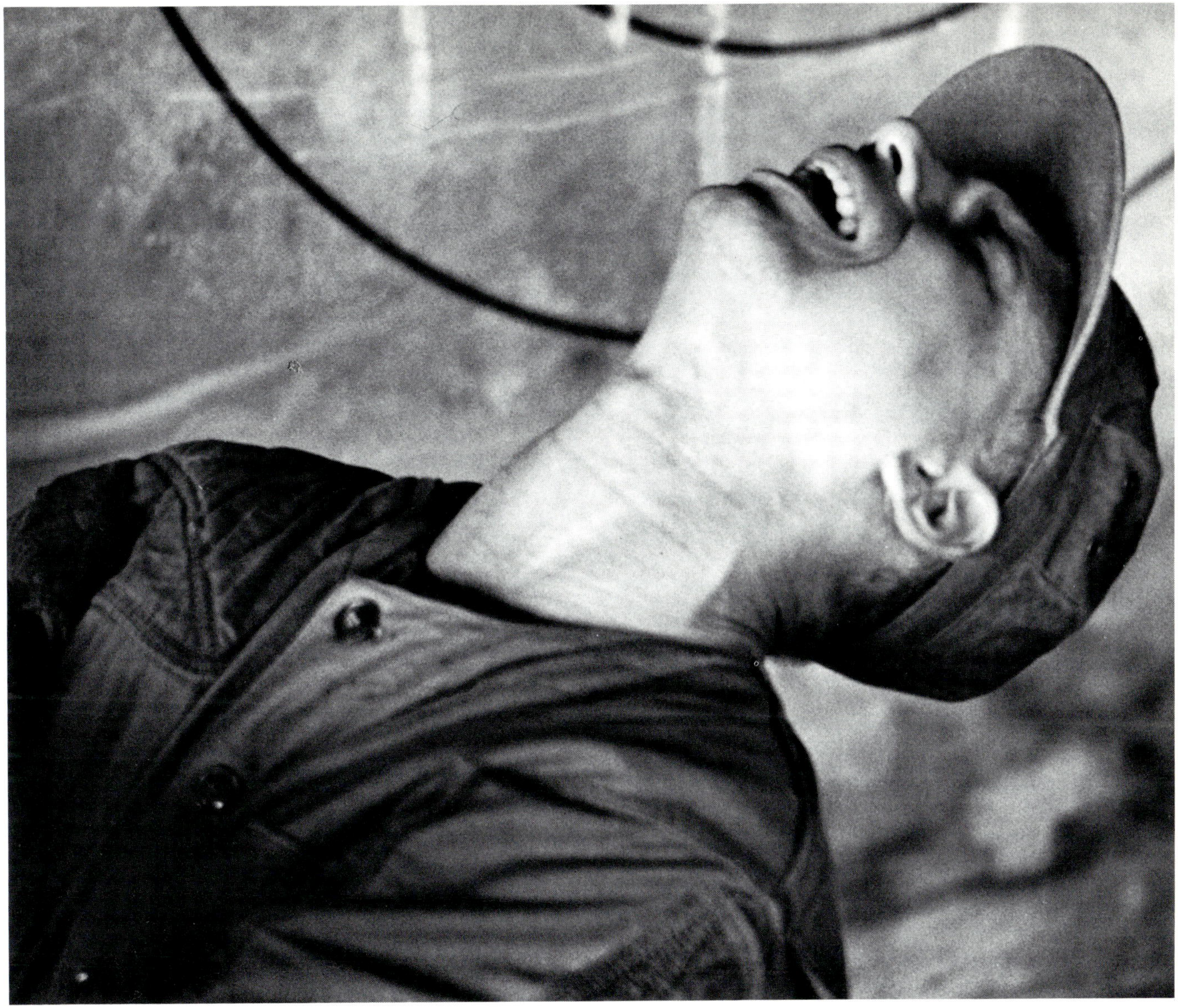

WHAT DO YOU THINK ABOUT PUGIL STICK PRACTICE?

"I like getting out there and try to knock someone's head off."

"Beat the hell out of him—KILL!! KILL!!!"

"Now this is what I wanted."

"Kill or be killed."

"Brutality with a laugh"

"Smack the guy in the head"

"Outstanding"

"Pure fun"

"Over come fear"

"Relieve tension"

"Scared of getting shit-kicked"

"Beat the fuck out of that pigfucker"

"Gave me a chance to prove myself to me"

"About time I get to let it out on somebody legally"

"I got the shit beat out of me."

—Responses to a written questionnaire

KNIFE ATTACK AGAINST A SENTRY:

a. Be noiseless.
b. Your attack must be launched not less than five feet from your opponent.
c. Your attack should be immediate and from your opponent's rear.
d. Thrust your knife upward, into his kidney.
e. Place your free hand over your opponent's mouth and nose.
f. Withdraw your knife with a ripping motion.
g. Lift your opponent's head and slash his jugular vein.

—The Red Book

NAKED STRANGLE HOLD: Stops the flow of blood from the heart to the brain; will cause unconsciousness in five to fifteen seconds. Death follows soon after unconsciousness if pressure is maintained.

a. Be noiseless.
b. Your attack should be immediate and from your opponent's rear.
c. Throw your right forearm across your opponent's neck while stepping forward with the right foot.
d. Move your left hand over your opponent's left shoulder and grasp your right hand—left palm up, right palm down.
e. Pull your opponent off balance, backwards to the deck.
f. Maintain a firm hold on your opponent as he drops to the deck. Simultaneously drop to your right knee and place your left foot flat on the deck.
g. Exert maximum constant pressure with a distinct sharp snap. Be certain the inner discs of your wrists are directed firmly against your opponent's carotid arteries.
h. Hold your position until you achieve the desired result.

—The Red Book

MILITARY ALIGNMENT

Trousers: The button-hole plait of the shirt is aligned with the fly of the trousers. The buckle is centered and on the same line.

—The Red Book

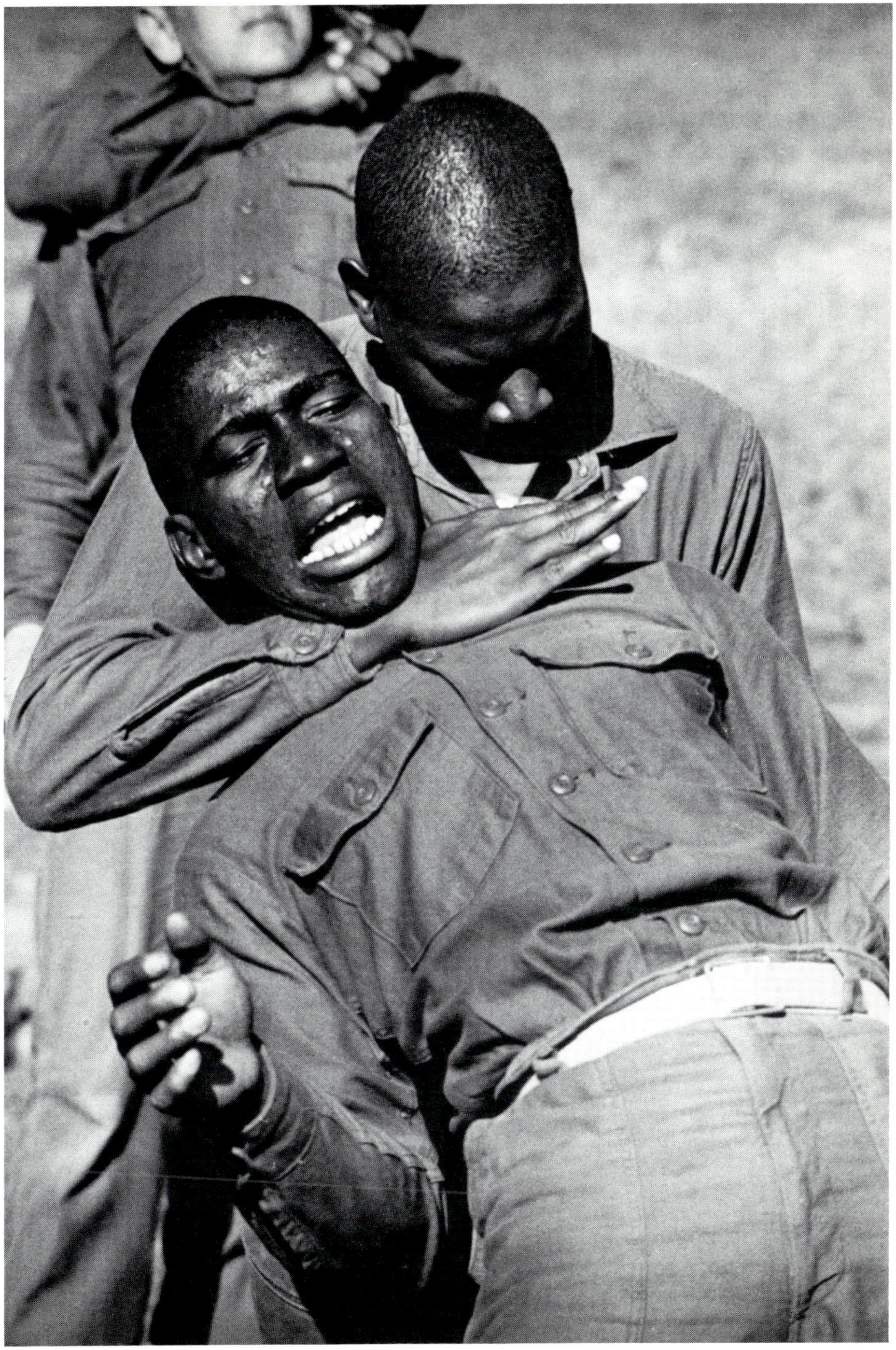

SERGEANT W. AT HOME

[Sergeant W. is a Senior Drill Instructor. On tape, he has an exceptionally mild and unassuming voice. It is hard to imagine his other personality on the drill field.]

INTERVIEWER: How does working eighteen hours a day running the recruits and then coming home work out? Do you automatically get out of the whole thing?

SERGEANT W.: I do. I don't know about anybody else, but I do. As soon as I get out of there I get away from the troops altogether. I think about the troops all the time, though. I don't like to be away from them too long because I don't know what's going on. I don't want anything to change, so I can't keep away from them too long. I sit around the house for two days and I have nothing to do, so I ride by the barracks. But the minute I get in or out of that car, I change. I don't come home and call cadence.
[Sergeant W. is interrupted by his small son, who comes in to inquire about a toolbox.]

SERGEANT W.: I don't want my kids doing it, but they do it anyway.

INTERVIEWER: Pardon me?

SERGEANT W.: My kids, they call cadence. I do say sometimes, "Give 'em doubletime." They know that means to run, so they start running down the road singing the Marine Corps Hymn. My daughter, she'll wander into the bedroom and start calling, "Left, Right, Lay-uft." They execute the hand salute.

INTERVIEWER: Do they ever watch you train the Recruits?"

SERGEANT W.: Oh, yeah. When we were on Mess and Maintenance my son would come into the barn. He likes to be near the Privates.

INTERVIEWER: How old is he?

SERGEANT W.: Seven. One time he asked me, "Do you still have that dumb recruit?" I asked him which one and he said, "You know, that one that don't know nothing." I said, "Yeah, I still have him." So when he came in the barn on Mess and Maintenance, he said, "Where is that dumb recruit that don't know anything?" The recruits were cleaning rifles. My son walked up to one and asked him what he was doing and the recruit said, "Cleaning my rifle, Sir." "Make sure you clean it good, Private," my son said.

INTERVIEWER: What did the Private say?

SERGEANT W.: "Yes, Sir."

—A taped interview

MOTIVATION COURSE

Observe these "do's" and "dont's":

(1) Flowers are not worn while in uniform

(2) Umbrellas are not carried while in uniform

—The Red Book

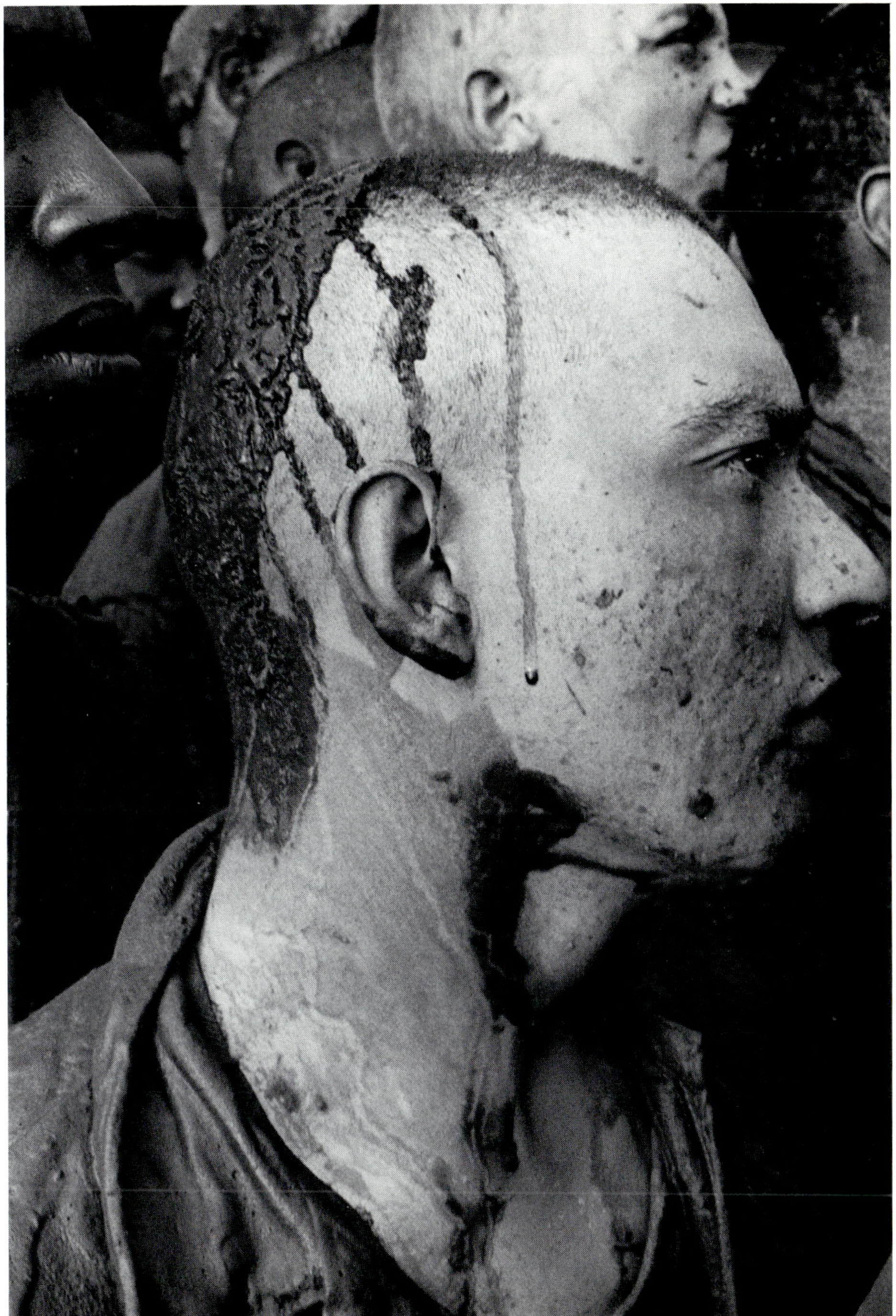

CARE OF THE FEET (IMPORTANT):

Wash your feet daily with soap and water. Dry them thoroughly after washing, especially between the toes, and dust generously with foot powder.

—The Red Book

[Drill Instructor, aside]: "They broke thirteen dummies—unrepairable. Those recruits were the meanest motherfuckers I seen in a long time."

"The dummies are on springs. They cost two hundred dollars apiece. Now's your time to get back at the Marine Corps—make them spend some money."

"Yes, Sir!"

"Now's the time to take out all your goddamned frustrations. At all times on that course I want to see just how goddamned nasty some of you fuckers are."

"Yes, SIR!"

"I want to see just how goddamned nasty and sadistic you can get on them goddamn dummies. If you can get nasty down there, I guarantee you can get nasty in combat. If you can't get nasty in practice, then you can't get nasty in the real thing, you understand?"

"YES, SIR!"

"I want to see you make a lot of goddamn noise. So when you hit that goddamn dummy, *hit* that motherfucker, you understand?"

"YES, SIR!"

"And if I see some of you cocksuckers fucking around, I'm going to take your weapon away, and guess what's going to happen? *You're* going to be the goddamn dummy!"

—Taped dialogue

WHAT DO YOU THINK ABOUT ALL THE YELLING YOU'RE SUPPOSED TO DO?

"Very motivating—KILL-KILL-KILL"

"Mind explosion"

"Aggressive"

"That's what we signed up for"

"Never knew I was allowed to yell stuff like that"

"Motervating"

"Opens your lungs and builds up your body to kill"

"Motivates the shit out of me"

"Repulsive"

"Sikes you up"

"Unbelievable feeling"

"Gives you a sore throat"

"Builds up confedence"

"Gives you strength"

"It's supposed to keep our minds off the real task—destroying life."

"I love it, makes me feel hateful."

"Blood tingling and exciting"

"Made me feel He man"

—Responses to a written questionnaire

"GUNG-HO"—Adopted from the Chinese and meaning "Working together."

—The Red Book

"Go ahead!"
"Hit him harder!"
"Hit him again!"
"Choke him!"
"CHOKE HIM!"
"CHOKE THAT SON OF A BITCH!"
"CHOKE HIM GOOD!"
"AHHHHHH."

—Taped dialogue

WHAT DOES GRADUATION MEAN TO YOU?

"Getting out of this hole in the ground"

"A great personal achievement"

"I made it."

"Changing from a boy to a man"

"The gift of God"

"I get to go home to my girl"

"I feel so tall I'm flying."

"Getting my hands on a beer"

"Getting off this fucking island"

"I deserve to be a Marine."

"It means I done it. I'm *proud.*"

"Just a ceremoney. More depending on what you make of it."

"A feeling of accomplishing something I hated so much."

"Escapeism, freedom"

"Seeing my family, BOOZE!! PUSSEY !!!!!"

"I've made it as a Marine. The best military personal in the world."

—Responses to a written questionnaire

WHO AND WHAT ARE THE MARINES?

"An elite group of men in a world of boys."

"The most efficient killing organization in the world—efficiency meaning killing more than you have killed."

"The guardians of U.S. and liberty."

"It's just people like you or me who've been made experts in aggressive assault."

"They're the only thing that could save this country with all the morals left out!"

"A team that works together better than any piece of machinery."

"Young men who care about their country."

"Nasty dudes that nobody fucks with."

"Men who protect God's country."

"The D.I., the Lord, and the best men in the world."

"The Marines are the men who make it."

"The Marine Corps is the best fighting unit in the world. They don't give a damn."

—Responses to a written questionnaire

About Harper Photographers

Harper Photographers is an experimental publishing program designed to encourage photographers to direct their thinking and their skills toward book-length projects dealing with a single theme. All books are in black and white, and the subject matter is completely open, as are the length and character of any text materials that may be used. The photographer may work alone, or in special cases, with up to two other people.

The primary goal of the program is to develop photography as the major element in original books, which must meet the same standards of creativity, style, cohesiveness of vision, and literary value as any other book. We regret that portfolios or photo-essay collections cannot be considered in the scope of the program.

Another goal of the Harper Photographers project is to make this work available to the public as inexpensively as possible All books will be printed by high-quality offset on medium-weight coated stock, and the best care will be given to the finest possible reproduction consistent with the books' low prices.

For further information, address Harper Photographers, 381/NJW, Harper & Row, Publishers, 10 East 53rd Street, New York, N.Y. 10022